Japan Math
Math Fact Mastery
Subtraction Facts

By Jody Weissler

Dear Parents and Teachers,

You made the right choice in purchasing this Japan Math workbook. Japan Math is a well-researched, theory based approach that leads to overall mathematical fluency. This book is one of a series of Japan Math books focused on math fact speed and retention. With practice and by following our guidelines, the Japan Math program has proven to lead children to unmatched fluency in their basic math facts.

We are proud that all of our Japan Math content is Common Core aligned and meets the mathematical standards of the 21st century.

In order to achieve success we strongly recommend that you follow the guidelines below.

- Copy the Student Data Record. This sheet shows all the math fact quizzes in this book and gives teachers and parents a place to record and monitor an individual student's academic progress.

- Don't make any assumptions. Start with the subtraction from 5 quiz.

- Consult the key in the back of the book (as needed) for scoring.

- Do not progress to the next quiz until mastery is reached on both forms.

- Use an online or traditional stop watch and note the time of each quiz (Always round up. Example: Two minutes should be circled if the quiz took 73 seconds)

- Mastery is 2 minutes or under and stick to that guideline.

- When the student has the mastered this entire book, it is time to move on to the next Japan Math workbook (addition, multiplication, division, or one of our word problem books).

Common Core Alignment

Grade Required Fluency
K Add and subtract within 5   (This book and our Addition Facts book)
1 Add and subtract within 10   (This book and our Addition Facts book)
2 Add and subtract within 20   (This book and our Addition Facts book)

Japan Math- Math Facts Student Data Record

Student's Name _____   Student's Grade Level   _____

Calendar Year _____   Mastery Date   _____

Subtraction Facts Timed Tests

Name of Test	Form	Date of Mastery	Form	Date of Mastery
Subtraction Facts Quiz Subtraction from 5	Form A		Form B	
Subtraction Facts Quiz Subtraction from 7	Form A		Form B	
Subtraction Facts Quiz Subtraction from 10	Form A		Form B	
Subtraction Facts Quiz Subtraction from 12	Form A		Form B	
Subtraction Facts Quiz Subtraction from 15	Form A		Form B	
Subtraction Facts Quiz Subtraction from 20	Form A		Form B	
Subtraction Facts Missing Part or Whole from 7	Form A		Form B	
Subtraction Facts Missing Part or Whole from 10	Form A		Form B	
Subtraction Facts Missing Part or Whole from 12	Form A		Form B	
Subtraction Facts Missing Part or Whole from 15	Form A		Form B	
Subtraction Facts Missing Part or Whole from 20	Form A		Form B	

Name_____   Date _____

Japan Math Subtraction Math Facts Mastery Quiz
(Subtraction from 5)

Time (in minutes)   2, 3, 4   Correct: _____/50
Japan Math Mastery = 50 Correct in Two Minutes!

5-3=	5-2=	5-4=	5-1=	5-3=
4-1=	4-3=	5-0=	5-2=	1-1=
4-3=	2-2=	5-3=	3-3=	5-1=
5-2=	5-0=	5-5=	5-4=	2-2=
5-1=	5-4=	3-2=	5-2=	4-2=
2-1=	5-3=	5-1=	5-4=	5-3=
5-3=	5-4=	3-3=	5-5=	2-1=
5-2=	1-1=	4-1=	3-2=	5-3=
4-3=	5-4=	3-1=	5-1=	4-3=
3-3=	4-3=	4-2=	5-5=	4-2=

More Japan Math Mastery at JapanMath.com

Name_____   Date _____

Form B

Japan Math Subtraction Math Facts Mastery Quiz
(Subtraction from 5)

Time (in minutes)   2, 3, 4   Correct: _____/50
Japan Math Mastery = 50 Correct in Two Minutes!

3-3=	3-2=	5-4=	4-1=	4-3=
4-1=	5-3=	3-1=	5-2=	1-1=
4-3=	2-1=	5-2=	3-3=	5-1=
5-3=	5-0=	5-5=	4-4=	2-2=
3-2=	5-4=	3-2=	5-4=	4-2=
2-1=	5-2=	5-1=	5-2=	4-3=
3-3=	5-4=	3-3=	5-5=	2-1=
5-2=	1-1=	5-1=	4-2=	4-3=
4-3=	5-4=	3-1=	5-1=	5-3=
3-3=	4-3=	4-2=	5-5=	5-4=

More Japan Math Mastery at JapanMath.com

Name_____   Date _____

Japan Math Subtraction Math Facts Mastery Quiz
(Subtraction from 7)

Time (in minutes)   2, 3, 4   Correct: ____/50
Japan Math Mastery = 50 Correct in Two Minutes!

7-6=	5-2=	5-5=	6-1=	4-3=
7-1=	5-3=	7-0=	6-2=	1-1=
7-3=	7-2=	6-3=	5-3=	5-1=
4-2=	7-0=	7-5=	7-4=	2-2=
2-2=	6-4=	3-2=	5-2=	4-2=
2-1=	7-5=	6-1=	7-4=	7-3=
6-3=	6-4=	3-3=	5-5=	2-1=
7-2=	1-1=	4-1=	7-6=	4-4=
4-3=	5-4=	3-1=	6-1=	6-3=
3-3=	7-3=	7-5=	5-5=	6-4=

More Japan Math Mastery at JapanMath.com

Name_____   Date _____

Form B

Japan Math Subtraction Math Facts Mastery Quiz
(Subtraction from 7)

Time (in minutes)   2, 3, 4   Correct: _____/50
Japan Math Mastery = 50 Correct in Two Minutes!

7-3=	7-2=	5-4=	5-1=	3-3=
6-1=	5-3=	6-0=	5-2=	6-1=
4-3=	6-2=	4-3=	5-3=	4-1=
7-2=	7-3=	7-4=	4-4=	2-2=
5-3=	7-4=	5-2=	4-2=	4-2=
3-2=	6-2=	4-2=	7-4=	6-3=
6-3=	6-5=	3-2=	6-4=	2-1=
7-4=	3-1=	3-1=	7-6=	5-4=
5-3=	6-4=	3-1=	6-2=	5-2=
4-3=	7-3=	6-5=	7-5=	6-4=

More Japan Math Mastery at JapanMath.com

Name_____   Date _____

Japan Math Subtraction Math Facts Mastery Quiz
(Subtraction from 10)

Time (in minutes)   2, 3, 4   Correct: _____/50
Japan Math Mastery = 50 Correct in Two Minutes!

7-6=	10-2=	10-9=	6-1=	4-3=
7-1=	5-3=	9-2=	6-2=	8-1=
7-3=	7-2=	6-3=	5-3=	8-6=
10-5=	10-7=	7-5=	10-4=	9-7=
9-8=	6-4=	8-5=	5-2=	4-2=
9-4=	7-5=	9-6=	10-4=	8-2=
8-3=	6-4=	10-1=	10-8=	9-5=
7-2=	9-1=	10-6=	10-3=	8-7=
10-5=	5-4=	10-7=	8-6=	9-3=
10-3=	8-4=	9-2=	9-6=	6-4=

More Japan Math Mastery at JapanMath.com

Name_____   Date_____

Form B

Japan Math Subtraction Math Facts Mastery Quiz
(Subtraction from 10)

Time (in minutes)   2, 3, 4   Correct: _____/50
Japan Math Mastery = 50 Correct in Two Minutes!

7-3=	10-2=	9-7=	5-1=	10-3=
7-4=	4-3=	8-2=	4-2=	5-2=
6-3=	4-2=	5-3=	8-3=	9-4=
10-5=	8-7=	7-5=	10-4=	9-7=
9-8=	6-3=	9-5=	5-2=	4-2=
9-4=	9-3=	9-6=	8-6=	7-3=
8-3=	6-4=	6-3=	7-4=	8-5=
10-2=	8-2=	8-3=	10-3=	9-5=
10-5=	5-4=	10-2=	8-6=	9-3=
6-3=	9-4=	9-2=	8-5=	6-4=

More Japan Math Mastery at JapanMath.com

Name_____   Date _____

Japan Math Subtraction Math Facts Mastery Quiz
(Subtraction from 12)

Time (in minutes)   2, 3, 4   Correct: _____/50
Japan Math Mastery = 50 Correct in Two Minutes!

7-4=	10-7=	6-5=	5-1=	12-3=
12-1=	5-3=	11-0=	6-2=	8-1=
8-3=	12-2=	11-5=	11-3=	5-1=
11-6=	7-0=	12-5=	10-4=	2-2=
10-8=	6-4=	11-2=	8-2=	10-2=
2-1=	7-5=	6-1=	12-4=	7-3=
12-4=	6-4=	3-3=	5-5=	12-8=
7-3=	9-1=	10-6=	8-3=	11-4=
12-7=	5-4=	12-1=	6-1=	9-3=
10-3=	7-3=	7-5=	12-5=	6-4=

More Japan Math Mastery at JapanMath.com

Name_____   Date _____

Form B

Japan Math Subtraction Math Facts Mastery Quiz
(Subtraction from 12)

Time (in minutes)   2, 3, 4   Correct: _____/50
Japan Math Mastery = 50 Correct in Two Minutes!

10-3=	7-3=	12-5=	8-1=	11-3=
11-6=	5-3=	11-6=	12-2=	8-3=
8-4=	6-2=	11-5=	11-3=	5-1=
11-4=	7-0=	12-5=	10-4=	12-4=
10-8=	6-4=	8-3=	6-2=	10-3=
9-3=	7-4=	11-7=	12-4=	8-3=
11-4=	6-4=	10-2=	5-5=	10-8=
7-3=	8-1=	8-6=	8-3=	9-4=
12-7=	5-4=	12-1=	11-2=	9-3=
10-3=	10-7=	11-5=	12-5	9-5=

More Japan Math Mastery at JapanMath.com

Name_____   Date_____

Japan Math Subtraction Math Facts Mastery Quiz
(Subtraction from 15)

Time (in minutes)   2, 3, 4   Correct: _____/50
Japan Math Mastery = 50 Correct in Two Minutes!

15-6=	10-2=	15-5=	14-9=	13-8=
14-7=	15-7=	13-9=	12-10=	13-5=
15-3=	12-9=	14-8=	13-6=	11-6=
12-8=	13-7=	15-8=	10-7=	12-7=
13-10=	15-9=	14-6=	11-8=	14-5=
12-9=	15-10=	9-6=	14-11=	13-9=
11-5=	12-7=	15-6=	10-8=	15-4=
14-9=	13-7=	10-6=	10-3=	11-4=
12-8=	11-9=	12-1=	15-7=	11-6=
12-3=	14-6=	15-12=	13-5=	14-8=

Name_____   Date_____

Form B

Japan Math Subtraction Math Facts Mastery Quiz
(Subtraction from 15)

Time (in minutes)   2, 3, 4   Correct: _____/50
Japan Math Mastery = 50 Correct in Two Minutes!

13-6=	10-1=	14-5=	14-7=	12-8=
12-7=	15-6=	12-9=	12-8=	13-5=
13-3=	14-9=	14-7=	13-6=	10-6=
15-8=	13-7=	15-8=	10-7=	12-7=
13-7=	15-4=	14-5=	11-9=	13-5=
12-9=	15-5=	10-6=	14-7=	15-9=
11-6=	12-7=	14-6=	10-8=	15-11=
14-10=	13-6=	11-5=	10-2=	11-4=
12-8=	11-9=	11-4=	15-7=	11-9=
12-4=	14-5=	13-8=	13-7=	14-8=

Name_____   Date_____

Japan Math Subtraction Math Facts Mastery Quiz
(Subtraction from 20)

Time (in minutes)   2, 3, 4   Correct: _____/50
Japan Math Mastery = 50 Correct in Two Minutes!

16-7=	15-6=	12-8=	16-9=	20-9=
13-7=	20-8=	18-4=	16-9=	19-5=
14-12=	15-3=	14-7=	14-6=	18-6=
14-6=	18-5=	20-6=	18-10=	16-13=
19-4=	14-9=	18-5=	18-8=	20-5=
19-9=	15-11=	17-11=	14-8=	17-9=
15-5=	18-13=	13-6=	16-8=	13-7=
18-9=	13-7=	18-4=	19-8=	14-2=
20-7=	20-8=	19-5=	15-7=	18-9=
16-2=	19-14=	17-12=	19-7=	20-6=

More Japan Math Mastery at JapanMath.com

Name_____   Date_____

Form B

Japan Math Subtraction Math Facts Mastery Quiz
(Subtraction from 20)

Time (in minutes) 2, 3, 4   Correct: _____/50
Japan Math Mastery = 50 Correct in Two Minutes!

12-8=	15-6=	14-8=	20-9=	16-9=
13-7=	17-8=	18-6=	19-9=	19-7=
20-12=	15-9=	16-7=	17-6=	18-7=
14-6=	19-5=	20-8=	18-11=	16-13=
17-4=	14-9=	20-6=	19-8=	14-5=
19-9=	15-10=	18-5=	14-11=	17-9=
20-5=	19-13=	15-6=	16-9=	15-8=
18-9=	13-7=	19-6=	17-8=	16-4=
18-7=	19-8=	17-6=	15-7=	20-11=
16-2=	19-14=	15-12=	19-5=	20-8=

More Japan Math Mastery at JapanMath.com

Name_____   Date _____

Japan Math Subtraction Math Facts Mastery Quiz
(Subtraction Facts Missing Part or Whole from 7)

Time (in minutes)   2, 3, 4   Correct: _____/50
Japan Math Mastery = 50 Correct in Two Minutes!

5- ☐ =1	☐ -2=3	☐ -5=0	6- ☐ =3	☐ -3=2
☐ -1=2	☐ -3=3	7- ☐ =3	6- ☐ =4	1- ☐ =1
7- ☐ =4	☐ -2=3	☐ -3=2	☐ -3=4	5- ☐ =2
☐ -2=5	☐ -3=4	☐ -5=2	☐ -4=3	2- ☐ =0
☐ -2=1	☐ -4=3	☐ -2=3	☐ -2=3	4- ☐ =1
2- ☐ =0	7- ☐ =3	6- ☐ =4	☐ -4=2	7- ☐ =3
☐ -3=4	6- ☐ =2	3- ☐ =1	☐ -5=2	☐ -1=3
7- ☐ =1	☐ -1=4	☐ -1=5	☐ -6=1	☐ -4=2
☐ -3=2	5- ☐ =3	3- ☐ =1	☐ -1=4	6- ☐ =3
3- ☐ =1	7- ☐ =3	☐ -5=1	☐ -5=2	6- ☐ =2

More Japan Math Mastery at JapanMath.com

Name_____   Date _____

Form B

Japan Math Subtraction Math Facts Mastery Quiz
(Subtraction Facts Missing Part or Whole from 7)

Time (in minutes)   2, 3, 4   Correct: _____/50
Japan Math Mastery = 50 Correct in Two Minutes!

□ - 3 = 1	7 - □ = 2	□ - 4 = 0	5 - □ = 4	□ - 3 = 4
□ - 1 = 6	□ - 3 = 3	7 - □ = 2	6 - □ = 3	1 - □ = 0
5 - □ = 3	□ - 4 = 3	□ - 3 = 3	□ - 3 = 4	5 - □ = 2
□ - 2 = 4	□ - 0 = 6	□ - 4 = 3	□ - 4 = 2	2 - □ = 1
□ - 2 = 5	□ - 4 = 3	□ - 2 = 4	□ - 2 = 3	4 - □ = 2
2 - □ = 0	6 - □ = 3	3 - □ = 2	□ - 4 = 3	7 - □ = 3
□ - 3 = 4	6 - □ = 2	7 - □ = 3	□ - 5 = 2	□ - 1 = 5
7 - □ = 6	□ - 1 = 3	□ - 7 = 0	□ - 6 = 1	□ - 4 = 3
□ - 3 = 4	7 - □ = 4	3 - □ = 1	□ - 5 = 1	6 - □ = 3
4 - □ = 2	6 - □ = 4	□ - 4 = 3	□ - 4 = 3	7 - □ = 2

More Japan Math Mastery at JapanMath.com

Name_____   Date _____

Japan Math Subtraction Math Facts Mastery Quiz
(Subtraction Facts Missing Part or Whole from 10)

Time (in minutes) 2, 3, 4   Correct: _____/50

Japan Math Mastery = 50 Correct in Two Minutes!

10-☐=3	9-☐=3	☐-3=6	6-☐=2	☐-4=5
☐-1=8	☐-3=4	7-☐=3	6-☐=3	2-☐=0
7-☐=2	☐-2=4	☐-3=7	☐-3=6	5-☐=4
☐-2=8	☐-0=10	☐-5=5	☐-4=4	10-☐=4
☐-3=2	☐-6=4	☐-2=8	☐-2=7	4-☐=1
8-☐=3	9-☐=4	10-☐=4	☐-2=4	10-☐=3
☐-3=7	6-☐=2	3-☐=1	☐-5=4	☐-1=7
7-☐=3	☐-1=6	☐-1=4	☐-6=3	☐-4=6
☐-3=6	5-☐=2	3-☐=1	☐-1=5	6-☐=3
7-☐=4	9-☐=6	☐-5=5	☐-5=3	9-☐=5

More Japan Math Mastery at JapanMath.com

Name_____   Date _____

Form B

Japan Math Subtraction Math Facts Mastery Quiz
(Subtraction Facts Missing Part or Whole from 10)

Time (in minutes)   2, 3, 4   Correct: _____/50
Japan Math Mastery = 50 Correct in Two Minutes!

☐-4=6	☐-2=3	☐-5=0	9-☐=2	☐-3=4
☐-1=8	☐-3=4	8-☐=5	10-☐=6	9-☐=5
7-☐=3	☐-2=8	☐-3=3	☐-4=4	7-☐=2
☐-2=8	☐-0=9	☐-5=2	☐-4=3	2-☐=1
☐-2=5	☐-4=6	☐-2=5	☐-2=6	4-☐=3
7-☐=3	7-☐=2	8-☐=3	☐-4=5	9-☐=2
☐-3=7	9-☐=2	9-☐=4	☐-5=5	☐-1=5
10-☐=2	☐-2=5	☐-1=3	☐-6=3	☐-4=5
☐-3=4	10-☐=3	9-☐=2	☐-1=5	6-☐=3
10-☐=6	7-☐=2	☐-5=4	☐-4=6	9-☐=5

More Japan Math Mastery at JapanMath.com

Japan Math Subtraction Math Facts Mastery Quiz
(Subtraction Facts Missing Part or Whole from 12)

Name_____   Date _____

Time (in minutes)   2, 3, 4   Correct: _____/50
Japan Math Mastery = 50 Correct in Two Minutes!

[ ]-4-=7	12-[ ]=3	[ ]-5=0	12-[ ]=3	[ ]-3=8
[ ]-3=8	[ ]-3=4	11-[ ]=6	11-[ ]=4	10-[ ]=2
7-[ ]=2	[ ]-2=10	[ ]-5=7	[ ]-3=5	8-[ ]=3
[ ]-2=5	[ ]-0=11	[ ]-5=6	[ ]-4=7	2-[ ]=1
[ ]-2=10	[ ]-4=7	[ ]-2=9	[ ]-2=4	4-[ ]=1
2-[ ]=1	11-[ ]=3	9-[ ]=5	[ ]-7=5	7-[ ]=3
[ ]-3=5	12-[ ]=3	11-[ ]=5	[ ]-4=2	[ ]-4=8
11-[ ]=	[ ]-1=11	[ ]-1=9	[ ]-6=5	[ ]-4=2
[ ]-3=8	9-[ ]=2	8-[ ]=3	[ ]-9=1	10-[ ]=7
3-[ ]=2	11-[ ]=3	[ ]-5=7	[ ]-7=4	12-[ ]=3

More Japan Math Mastery at JapanMath.com

Name_____   Date _____

Form B

Japan Math Subtraction Math Facts Mastery Quiz
(Subtraction Facts Missing Part or Whole from 12)

Time (in minutes) 2, 3, 4   Correct: _____/50
Japan Math Mastery = 50 Correct in Two Minutes!

11- ☐ =3	11- ☐ =5	☐ -7=3	11- ☐ =9	☐ -3=7
☐ -1=8	☐ -3=8	12- ☐ =8	9- ☐ =3	10- ☐ =3
10- ☐ =6	☐ -2=10	☐ -3=9	☐ -3=8	5- ☐ =3
☐ -2=8	☐ -0=11	☐ -5=7	☐ -4=5	10- ☐ =6
☐ -2=10	☐ -4=7	☐ -2=9	☐ -2=9	11- ☐ =7
12- ☐ =5	10- ☐ =3	9- ☐ =6	☐ -4=8	7- ☐ =2
☐ -3=1	11- ☐ =4	12- ☐ =3	☐ -5=4	☐ -1=10
12- ☐ =3	☐ -1=9	☐ -1=11	☐ -4=2	12- ☐ =5
☐ -3=7	9- ☐ =3	3- ☐ =1	☐ -1=4	9- ☐ =2
12- ☐ =	7- ☐ =4	☐ -3=8	☐ -5=7	11- ☐ =4

More Japan Math Mastery at JapanMath.com

Name_____   Date _____

Japan Math Subtraction Math Facts Mastery Quiz
(Subtraction Facts Missing Part or Whole from 15)

Time (in minutes)   2, 3, 4   Correct: _____/50
Japan Math Mastery = 50 Correct in Two Minutes!

☐-9=6	15-☐=3	☐-5=6	14-☐=11	☐-3=10
☐-8=7	☐-3=12	11-☐=3	12-☐=5	14-☐=8
15-☐=12	☐-12=3	☐-3=12	☐-3=11	15-☐=2
☐-4=11	☐-2=13	☐-5=7	☐-4=11	12-☐=1
☐-2=12	☐-4=11	☐-2=12	☐-2=13	4-☐=1
14-☐=3	13-☐=7	12-☐=5	☐-4=5	12-☐=3
☐-3=12	14-☐=7	13-☐=7	☐-5=9	☐-5=10
7-☐=2	☐-7=8	☐-6=6	☐-6=9	☐-4=8
☐-3=10	15-☐=9	13-☐=6	☐-11=4	12-☐=8
13-☐=9	12-☐=8	☐-5=9	☐-5=6	12-☐=5

More Japan Math Mastery at JapanMath.com

Name_____   Date _____

Form B

Japan Math Subtraction Math Facts Mastery Quiz
(Subtraction Facts Missing Part or Whole from 15)

Time (in minutes)   2, 3, 4   Correct: _____/50
Japan Math Mastery = 50 Correct in Two Minutes!

$13-\boxed{}=1$	$14-\boxed{}=3$	$\boxed{}-5=0$	$6-\boxed{}=2$	$\boxed{}-3=12$
$\boxed{}-2=12$	$\boxed{}-3=12$	$7-\boxed{}=2$	$15-\boxed{}=6$	$11-\boxed{}=7$
$7-\boxed{}=3$	$\boxed{}-10=3$	$\boxed{}-3=12$	$\boxed{}-3=11$	$15-\boxed{}=5$
$\boxed{}-6=9$	$\boxed{}-2=11$	$\boxed{}-5=9$	$\boxed{}-4=8$	$12-\boxed{}=5$
$\boxed{}-3=12$	$\boxed{}-4=11$	$\boxed{}-2=8$	$\boxed{}-2=11$	$14-\boxed{}=8$
$12-\boxed{}=6$	$13-\boxed{}=7$	$14-\boxed{}=8$	$\boxed{}-4=11$	$15-\boxed{}=2$
$\boxed{}-3=12$	$14-\boxed{}=9$	$13-\boxed{}=6$	$\boxed{}-5=10$	$\boxed{}-4=11$
$14-\boxed{}=6$	$\boxed{}-8=7$	$\boxed{}-6=9$	$\boxed{}-8=6$	$\boxed{}-5=8$
$\boxed{}-3=12$	$15-\boxed{}=4$	$15-\boxed{}=6$	$\boxed{}-7=7$	$12-\boxed{}=3$
$13-\boxed{}=4$	$15-\boxed{}=8$	$\boxed{}-12=3$	$\boxed{}-8=7$	$11-\boxed{}=5$

More Japan Math Mastery at JapanMath.com

Name_____   Date _____

Japan Math Subtraction Math Facts Mastery Quiz
(Subtraction Facts Missing Part or Whole from 20)

Time (in minutes) 2, 3, 4   Correct: _____/50
Japan Math Mastery = 50 Correct in Two Minutes!

☐ $-8=1$	$18-$ ☐ $=3$	☐ $-5=14$	$16-$ ☐ $=8$	☐ $-13=6$
☐ $-4=16$	☐ $-3=16$	$17-$ ☐ $=12$	$18-$ ☐ $=13$	$15-$ ☐ $=6$
$17-$ ☐ $=8$	☐ $-3=17$	☐ $-4=15$	☐ $-6=8$	$15-$ ☐ $=7$
☐ $-4=15$	☐ $-3=12$	☐ $-13=7$	☐ $-6=14$	$12-$ ☐ $=4$
☐ $-6=13$	☐ $-3=9$	☐ $-12=8$	☐ $-8=8$	$14-$ ☐ $=9$
$17-$ ☐ $=9$	$17-$ ☐ $=11$	$16-$ ☐ $=12$	☐ $-14=5$	$17-$ ☐ $=4$
☐ $-3=12$	$16-$ ☐ $=9$	$13-$ ☐ $=9$	☐ $-5=7$	☐ $-4=8$
$17-$ ☐ $=7$	☐ $-12=8$	☐ $-3=12$	☐ $-6=12$	☐ $-4=7$
☐ $-3=9$	$15-$ ☐ $=2$	$20-$ ☐ $=4$	☐ $-11=9$	$16-$ ☐ $=12$
$13-$ ☐ $=7$	$17-$ ☐ $=12$	☐ $-5=11$	☐ $-5=8$	$16-$ ☐ $=5$

More Japan Math Mastery at JapanMath.com

Name_____   Date _____

Form B

Japan Math Subtraction Math Facts Mastery Quiz
(Subtraction Facts Missing Part or Whole from 20)

Time (in minutes)   2, 3, 4   Correct: _____/50
Japan Math Mastery = 50 Correct in Two Minutes!

☐-3-=17	20-☐=16	☐-5=0	16-☐=7	☐-3=16
☐-1=14	☐-9=7	17-☐=8	14-☐=10	15-☐=4
18-☐=7	☐-4=3	☐-9=9	☐-11=3	19-☐=11
☐-17=2	☐-9=5	☐-8=7	☐-12=6	20-☐=3
☐-5=7	☐-11=7	☐-8=9	☐-9=11	18-☐=12
14-☐=8	17-☐=5	20-☐=14	☐-13=6	17-☐=13
☐-13=6	16-☐=8	13-☐=8	☐-6=4	☐-7=12
17-☐=9	☐-7=12	☐-5=12	☐-6=8	☐-4=13
☐-5=1	9-☐=2	13-☐=6	☐-11=8	16-☐=9
13-☐=5	17-☐=14	☐-6=13	☐-5=5	17-☐=9

More Japan Math Mastery at JapanMath.com

Answer Keys to Japan Math Subtraction Facts Assessments

Subtraction Facts from 5

2	3	1	4	2
3	1	5	3	0
1	0	2	0	4
3	5	0	1	0
4	1	1	3	2
1	2	4	1	2
2	1	0	0	1
3	0	3	1	2
1	1	2	4	1
0	1	2	0	2

Subtraction Facts from 5 (Form b)

0	1	1	3	1
3	2	2	3	0
1	1	3	0	4
2	5	0	0	0
1	1	1	1	2
1	3	4	3	1
0	1	0	0	1
3	0	4	2	1
1	1	2	4	2
0	1	2	0	1

Subtraction Facts from 7

1	3	0	5	1
6	2	7	4	0
4	5	3	2	4
2	7	2	3	0
0	2	1	3	2
1	2	5	3	4
3	2	0	0	1
5	0	3	1	0
1	1	2	5	3
0	4	2	0	2

Subtraction Facts from 7 (Form b)

4	5	1	4	0
5	2	6	3	5
1	4	1	1	3
5	4	3	0	0
2	3	3	2	2
1	4	2	3	3
3	1	1	2	1
3	2	2	1	1
2	2	2	4	3
1	4	1	2	2

Subtraction Facts from 10

1	8	1	5	1
6	2	7	4	7
4	5	3	2	2
5	3	2	6	2
1	2	3	3	2
5	2	3	6	6
5	2	9	2	4
5	8	4	7	1
5	1	3	2	6
7	4	7	3	2

Subtractin Facts from 10 (Form b)

4	8	2	4	7
3	1	6	2	3
3	2	2	5	5
5	1	2	6	2
1	3	4	3	2
5	6	3	2	4
5	2	3	3	3
8	6	5	7	4
5	1	8	2	6
3	5	7	3	2

Subtraction Facts From 12

3	33	1	4	9
11	2	11	4	7
5	10	6	8	4
5	7	7	7	0
2	2	9	6	8
1	2	5	8	4
8	2	0	0	3
4	8	4	5	7
5	1	11	5	6
7	4	2	7	2

Subtraction Facts From 12 (Form b)

7	4	7	7	8
5	2	5	10	5
4	4	6	8	4
7	7	7	6	8
2	2	5	4	7
6	3	4	8	5
7	2	8	0	5
4	7	2	5	5
5	1	11	9	6
7	3	6	7	4

Subtracton Facts Sum to 15

9	8	10	5	5
7	8	4	2	8
12	3	6	7	5
4	6	7	3	5
3	6	8	3	9
3	5	3	3	4
6	5	9	2	11
5	6	4	7	7
4	2	11	8	5
9	8	3	8	6

Subtraction Facts from 15 (Form b)

7	9	9	7	4
5	9	3	4	8
10	5	7	7	4
7	6	7	3	5
6	11	9	2	8
3	10	4	7	6
5	5	8	2	4
4	7	6	8	7
4	2	7	8	2
8	9	5	6	6

Subtraction Facts from 20

9	9	4	7	11
6	12	14	7	14
2	12	7	8	12
8	13	14	8	3
15	5	13	10	15
10	4	6	6	8
10	5	7	8	6
9	6	14	11	2
13	12	14	8	9
14	5	5	12	14

Subtraction Facts from 20 (Form b)

4	9	6	11	7
6	9	12	10	12
8	6	9	11	11
8	14	12	7	3
13	5	14	11	9
10	5	13	3	8
15	6	9	7	7
9	6	13	9	12
11	11	11	8	9
14	5	3	14	12

Subtraction Facts Missing Part or Whole to 7

4	5	5	3	5
3	6	4	2	0
3	5	5	7	3
7	7	7	7	2
3	7	5	5	3
0	4	2	6	4
7	4	2	7	4
6	5	6	7	6
5	2	2	5	3
2	4	6	7	4

Subtraction Facts Missing Part or Whole to 7 (form b)

4	5	4	1	7
7	6	5	3	1
2	7	6	7	3
6	6	7	6	1
7	7	6	5	2
2	3	1	7	4
7	4	4	7	6
1	4	7	7	7
7	3	2	6	3
2	2	7	7	5

Subtraction Facts Missing Part or Whole to 10

7	6	9	4	9
9	7	4	3	2
5	6	10	9	1
10	10	10	8	6
5	10	10	9	3
5	5	6	6	7
10	4	2	9	8
4	7	5	9	10
9	3	2	6	3
3	3	10	8	4

Subtraction Facts Missing Part or Whole to 10 (form b)

10	5	5	7	7
9	7	3	4	4
4	10	3	4	5
6	9	7	7	1
7	10	7	8	1
4	5	5	9	7
10	7	5	10	6
8	7	4	9	9
7	7	7	6	3
6	5	9	10	4

Subtraction Facts Missing Part or Whole to 12

11	9	5	9	11
11	7	5	7	8
5	12	12	8	5
7	11	11	11	1
12	11	11	6	3
1	8	4	12	4
8	9	6	6	12
5	12	10	11	6
11	7	5	10	3
1	8	12	11	9

Subtraction Facts Missing Part or Whole to 12 (form b)

8	6	10	2	10
9	11	4	6	7
4	12	12	11	2
10	11	12	9	4
12	11	11	11	4
7	7	3	12	5
4	7	9	9	11
9	10	12	6	7
10	6	2	5	7
5	3	11	12	7

Subtraction Facts Missing Part or Whole to 15

15	12	11	3	13
15	15	8	7	6
3	15	15	14	13
15	15	12	15	11
14	15	14	15	3
11	6	7	9	9
15	7	6	14	15
5	15	12	15	12
10	6	7	15	4
4	4	14	11	7

Subtraction Facts Missing Part or Whole to 15 (form b)

12	11	5	4	15
14	15	5	9	4
4	13	15	14	10
15	13	14	12	7
15	15	10	13	6
6	6	6	15	13
15	5	7	15	15
8	15	15	14	13
15	11	9	14	9
9	7	15	15	6

Subtraction Facts Missing Part or Whole to 20

9	15	19	8	19
20	19	5	5	9
9	20	19	14	8
19	15	20	20	8
19	12	20	16	5
8	6	4	19	13
15	7	4	12	8
10	20	15	18	11
12	13	16	20	4
6	5	16	13	11

Subtraction Facts Missing Part or Whole to 20 (form b)

20	4	5	9	19
15	16	9	4	11
11	7	18	14	8
19	14	15	18	17
12	18	17	20	16
6	12	6	19	4
19	8	5	10	19
8	19	17	14	17
6	7	7	19	7
8	3	19	10	12